11th Workshop on Principles and Applications of Control in Quantum Systems (PRACQSYS 2017)

AA005222

Seattle, Washington, USA
17 - 20 July 2017

ISBN: 978-1-5108-5020-0

Printed from e-media with permission by:

Curran Associates, Inc.
57 Morehouse Lane
Red Hook, NY 12571

Some format issues inherent in the e-media version may also appear in this print version.

This material is available in the "Public Domain" and can be downloaded at the following link:
http://pracqsys2017.ee.washington.edu/files/2017/08/PRACQSYS2017_Proceedings.pdf

Printed by Curran Associates, Inc. (2018)

For additional information, please contact University of Washington/Department of Electrical Engineering at the address below.

University of Washington/Department of Electrical Engineering
Campus Box 352500
185 Stevens Way
Paul Allen Center
Seattle, Washington 98195-2500
USA

Phone: 650-906-8666

help@uw.edu

Additional copies of this publication are available from:

Curran Associates, Inc.
57 Morehouse Lane
Red Hook, NY 12571 USA
Phone: 845-758-0400
Fax: 845-758-2633
Email: curran@proceedings.com
Web: www.proceedings.com

TABLE OF CONTENTS

PRACQSYS 2017

Principles and Applications of Control in Quantum Systems

11th Workshop

Conference Proceeding

(http://pracqsys2017.ee.washington.edu/)

ELECTRICAL ENGINEERING
UNIVERSITY of WASHINGTON

DEPARTMENT OF PHYSICS
UNIVERSITY of WASHINGTON

PAUL G. ALLEN SCHOOL
OF COMPUTER SCIENCE & ENGINEERING

SCIENCE TO SHAPE THE FUTURE OF THE ARMY
ARMY RESEARCH OFFICE
1951

NSF

Microsoft

Monday, July 17th

8:00	Registration +Breakfast
8:45	Welcome Remark
Session Chair:	Arka Majumdar
9:00	*Optical Neural Networks at Quantum Limit*, Yoshi Yamamoto, ImPACT Program, Japan Science and Technology Agency
9:40	*The Power of Being Positive: Compressed Sensing in the Quantum World*, Ivan Deutsch, University of New Mexico
10:20	Coffee Break
11:00	*Quantum Control in quantum metrology*, Haidong Yuan, Chinese University of Hong Kong
11:40	Lunch (on your own)
Session Chair:	Ivan Deutsch
1:30	*Control and Readout of Semiconductor Double Quantum Dot Qubits using Circuit Quantum Electrodynamics (QED)*, Andreas Wallraff, ETH Zurich
2:10	*Quantum control of ion-photon interactions in an optical cavity*, Tracy Northup, UIBK
2:50	Coffee Break
3:30	*Identification and estimation in quantum input-output systems*, Madalin Guta, University of Nottingham
4:10	*Quantum Control of Superconducting Circuits*, Victor Albert, Yale University

Tuesday, July 18th

8:00	Registration +Breakfast
Session Chair:	Michael J. Biercuk
9:00	*Controlled dynamics of superconducting qubits for quantum chemistry calculations*, Stefan Filipp, IBM
9:40	*All-optical Computational Circuits in Integrated Photonics*, Raymond G Beausoleil, HP Labs

10:20	Coffee Break
11:00	*Layout and Control for Measurement Based Quantum Computing using Topological Qubits*, Krysta Svore, Microsoft
11:40	Lunch (on your own)
Session Chair:	Kai-Mei Fu
1:45	Discussion forum: role of industry in quantum control (Moderator: Kai-Mei Fu; Group leaders: Stefan Filipp, Raymond G Beausoleil, and Michael J. Biercuk)
2:30	Coffee Break
3:00	Informal report out of the discussion forum
3:30	*Tomography of an Optomechanical Oscillator via Position Measurement*, Andrew Doherty, University of Sydney
4:10	*Quantum Noise Spectroscopy via Open-Loop Control on Qubit Sensors*, Lorenza Viola, Dartmouth College
5:00	Poster session with refreshments (till 6:30 pm)

Wednesday July 19th

8:00	Registration +Breakfast
Session Chair:	Francisco Elohim Becerra
9:00	*Quantum input-output networks in silicon photonics: prospects and applications*, Mohan Sarovar, Sandia Labs
9:40	*Quantum nanophotonics: from inverse design to applications*, Jelena Vuckovic, Stanford University
10:20	Coffee Break
11:00	Lunch (on your own)
Session Chair:	Arka Majumdar
1:30	Discussion forum: integrated photonics for quantum control (Moderator: Arka Majumdar; group leaders: Kai-Mei Fu, Mohan Sarovar, Jelena Vuckovic and Raymond G Beausoleil)
2:30	Coffee break
3:00	Informal report out of the discussion forum

3:30 *Optimized strategies for nonorthogonal state discrimination at the single-photon level*, Francisco Elohim Becerra, University of New Mexico

4:10 *Control of Quantum Spin Devices, feedback control laws and hidden feedback,* Sophie Schirmer, Swansea University

5:30-7:30 Buffet Dinner

Thursday July 20th

8:00 Registration +Breakfast

Session Chair: Klemens Hammerer

9:00 *Parity-time symmetry and chaos transfer in optomechanics*, Jing Zhang, Tsinghua University

9:40 *Photonic controlled-phase gates through Rydberg blockade in optical cavities*, Etienne Brion, Laboratoire Aimé Cotton

10:20 Coffee Break

11:00 *Oscillators for quantum information: the cat code, error correction, and two mode entanglement*, Robert Schoelkopf, Yale University

11:40 *The Magic of Combining Switchable Noise with Coherent Control: Principles and Implementation in Superconducting Qubits*, Thomas Schulte-Herbrueggen, Technical University Munich

12:20 Lunch (on your own)

Session Chair: Nathan Weibe

1:30 *Beyond Markovian approximations in open quantum systems*, Alain Sarlette, Inria

2:10 *Echoes of Entanglement: from Quantum Metrology to Scrambling*, Monika Schleier-Smith, Stanford University

2:50 *Quantum Control Exploiting Light-Matter Entanglement*, Klemens Hammerer, Leibniz University

3:30 Closing of the conference; committee (local + IPC) meets

Talk Format: 35 minutes presentation + 5 minutes Q&A

Title: The Magic of Combining Switchable Noise with Coherent Control: Principles and Implementation in Superconducting Qubits

Thomas Schulte-Herbrueggen
Technical University Munich
Email: tosh@tum.de

Abstract: Combining coherent control with simplest noise switching seems magic: it allows to interconvert arbitrary n-qubit quantum states no matter whether they are pure or mixed. We analyze switchable noise at the limits of open-loop control in view of principle limits between Markovian and non-Markovian control (realized by closed-loop feedback). We suggest experimental implementation in superconducting GMon devices via our numerical platform DYNAMO. All these findings fit nicely in a Lie-geometric picture of dynamic systems control.

Title: Beyond Markovian approximations in open quantum systems

Alain Sarlette
INRIA (QUANTIC) & Ghent University (SYSTeMS)
Email: alain.sarlette@inria.fr

Abstract: In this talk we will review our recent results about analyzing the effect of time correlations in open quantum systems from various viewpoints. These include (i) the effect of entanglement over time, in the input to engineered reservoirs; (ii) applications of our recently developed systematic method for adiabatic elimination of fast dynamics, to include dynamic effects into fast converging systems; and possibly (iii) adding simple dynamics into quantum feedback loops (i.e. extending so-called Markovian feedback towards standard control theory elements).

Title: Layout and Control for Measurement Based Quantum Computing using Topological Qubits

Krysta Svore
Station Q, Microsoft Research
Email: ksvore@microsoft.com

Abstract: Microsoft's Station Q was founded in 2006. The goal of the team has always been a topological quantum computer. Recent breakthroughs have led us to a Measurement Based architecture (see: https://arxiv.org/abs/1610.05289). This talk will focus on the motivation, challenges and solutions being addressed. We will present the Microsoft $Sol\,|i\rangle$ compiler and runtime as the vehicle to achieve our goal of operating a Topological Quantum Computer as well as the challenges in building a scalable architecture.

Title: The Power of Being Positive: Compressed Sensing in the Quantum World

Ivan Deutsch
University of New Mexico
Email: ideutsch@unm.edu

Abstract: Diagnosing the performance of quantum information processors is a grand challenge. This is because quantum information is stored in a manner quite different from classical information. We can't just look under the hood and measure everything we want about the device in one fell swoop. Quantum measurements disturb the system and we can't learn everything about non-commuting observables at the same time. This becomes particularly challenging for systems that encode information in a large dimensional Hilbert space, e.g. $d = 2^N$, for N qubits. Thus, we must devise special methods for characterizing, verifying, and validating the device performance. One tool we can borrow from classical estimation theory is "compressed sensing," which employs prior information about the structure of a signal to greatly reduce the resources we need to reconstruct it. In this talk, I will show how compressed sensing follows almost automatically in estimation of quantum states and processes due to fact that the physics imposes a constraint — they are described by positive matrices. This important geometrical feature gives great flexibility in implementing compressed-sensing tomography of states and processes. I will describe the fundamental theory of compressed sensing tomography and its implementation in an experimental testbed consisting of hyperfine spins in ultracold cesium atoms, defined by a 16-dimensional Hilbert space. Experimental work has been carried out by my collaborator, Prof. Poul Jessen, University of Arizona.

Title: Controlled dynamics of superconducting qubits for quantum chemistry calculations

Stefan Filipp
IBM Research Zurich, 8803 Rueschlikon, Switzerland
Email: ZRLSFI@ch.ibm.com

Abstract: Precise control of individual qubits and their interactions is essential for quantum information processing applications, which offer the potential to surpass limits of classical computability. While the computation of molecular energy spectra is a hard task on a classical computer, it may benefit from the possibility to efficiently implement entangled multi-qubit states on a quantum platform. In this talk, I will present recent experiments mapping out the ground state energies of simple molecules using a variational method on a multi-qubit quantum processor. In this approach, the ground state is prepared by a sequence of single-qubit and microwave-induced entangling gates acting on fixed-frequency transmon qubits; the gate parameters resulting in the ground-state wavefunctions are found by a classical algorithm minimizing the energy as the cost function. Moreover, I will discuss alternative methods to control the interactions between two qubits based on parametric driving of a frequency-tunable coupler, a scheme suited to engineer Hamiltonians with coupling terms adjustable in strength, phase and orientation. Aside from providing an alternative two-qubit gate mechanism, this method can be used to adiabatically steer the system from a trivial, separable initial state to an entangled final state encoding the ground state of a molecular system.

Title: Optical Neural Networks at Quantum Limit

Yoshihisa Yamamoto
ImPACT Program, Japan Science and Technology Agency, 7 Gobancho, Chiyoda-ku, Tokyo 102-0076, JAPAN
E. L. Ginzton Laboratory, Stanford University, Stanford, California 94305, USA
Email: yyamamoto@stanford.edu

Abstract: We describe the basic concept and operational principle of a novel computing machine, optical neural networks operating at quantum limit, in this talk. Degenerate optical parametric oscillator pulses in a fiber ring cavity are used as quantum neurons, while optical homodyne detectors and PFGA feedback circuits are employed as quantum synapses. Two types of optical neural networks are introduced: symmetric neural networks for solving NP-hard Ising problems and asymmetric recurrent neural network for solving NP-complete k-SAT problems. We also discuss the performance of two experimental systems: one is Stanford machine with 100 neurons and 10^4 synapses and the other is NTT machine with 2000 neurons and 4 x 10^6 synapses.

Title: Optimized strategies for nonorthogonal state discrimination at the single-photon level

Francisco Elohim Becerra
University of New Mexico
Email: fbecerra@unm.edu

Abstract: The discrimination of nonorthogonal states is not possible to perform with total certainty by any deterministic measurement. This makes nonorthogonal states, such as coherent states of light, a tool to achieve secure communication via quantum key distribution. However, the intrinsic nonorthogonality of coherent states also sets fundamental limits on the amount of information that can be communicated in low-power optical communication. Here, I describe optimized discrimination strategies for multiple coherent states based on adaptive measurements with optimized displacement operations and photon counting, which can surpass the conventional limit of coherent detection, called the quantum noise limit (QNL), at arbitrary input power levels. We use these strategies to demonstrate the discrimination of multiple coherent states below the QNL at the single-photon level under realistic conditions with loss and noise. These measurement strategies can potentially enhance information transfer beyond what can be achieved with optical heterodyne detection and multiple coherent states.

Title: Quantum input-output networks in silicon photonics: prospects and applications

Mohan Sarovar
Sandia Labs
Email: mnsarov@sandia.gov

Abstract: Quantum input-output networks (QIONs) and associated modeling tools, including the SLH framework, show great promise for realizing quantum information processing and communication primitives. We have recently studied the applicability of the integrated silicon photonics platform for implementing scalable QIONs. I will discuss the strengths of the silicon photonics platform for quantum applications and also talk about some of the key challenges to both the theoretical formalism and experimental implementation of QIONs. In particular, I will discuss some of the extensions to SLH framework (which was originally developed with bulk-optics implementations in mind) required to make it fully applicable to modeling of linear and nonlinear integrated optics networks.

Title: Quantum Control of Superconducting Circuits

Liang Jiang
Yale University
Email: liang.jiang@yale.edu

Abstract: We have developed an efficient quantum control scheme that allows for arbitrary unitary operations on a cavity mode using strongly dispersive qubit-cavity interaction and time-dependent drives [1,2]. Combined with measurement and adaptive control, we can implement arbitrary completely positive and trace preserving (CPTP) maps over the cavity mode [3]. Our control scheme can be implemented using superconducting circuit, demonstrating quantum error correction [4] and quantum logical gates [5]. Moreover, we may also engineer the dissipation to achieve holonomic quantum computation [6,7].

[1] Krastanov, et al., PRA 92, 040303 (2015)
[2] Heeres, et al., PRL 115, 137002 (2015)
[3] Shen, et al., PRB 95, 134501 (2017)
[4] Ofek, et al., Nature 536, 441 (2016)
[5] Heeres, et al., Nature Communications (accepted); arXiv:1608.02430.
[6] Albert, et al., PRL 116, 140502 (2016)
[7] Albert, et al., PRX 6, 041031 (2016)

Title: Quantum nanophotonics: from inverse design to applications

Jelena Vuckovic
Stanford University
Email: jela@stanford.edu

Abstract: Nanophotonic structures that localize photons in sub-wavelength volumes are possible today thanks to modern nanofabrication and optical design techniques. Such structures enable studies of new regimes of light-matter interaction, quantum and nonlinear optics, and new applications in computing, communications, and sensing. The traditional quantum nanophotonics platform is based on InAs quantum dots inside GaAs photonic crystal cavities [1]. Recently, alternative material systems have emerged, such as color centers in diamond and silicon carbide, that could potentially bring the described experiments to room temperature and facilitate scaling to large networks of resonators and emitters [2-3]. Finally, the use of inverse design nanophotonic methods that can efficiently perform physics-guided search through the full parameter space, leads optical devices with properties superior to state of the art, including smaller footprints, better field localization, and novel functionalities [4].

1. *Nature Photonics,*vol. 10, pp. 163-166 (2016)
2. *Nano Letters, vol.* 16 (1), pp. 212-217 (2016)
3. *Nano Letters, vol.* 17 (3), pp 1782–1786 (2017)
4. *Nature Photonics* 9, 374–377 (2015)

Title: Quantum Control Exploiting Light-Matter Entanglement

Klemens Hammerer
Leibniz Universität Hannover
Email: Klemens.Hammerer@itp.uni-hannover.de

Abstract: The generation of quantum states carrying entanglement among propagating light fields and stationary matter is a prerequisite for fundamental test of quantum mechanics, such as loophole free Bell tests, as well as for applications in quantum communication over long distances. Current experiments achieve a remarkably high efficiency in generating and controlling such entangled states of matter, such as single atoms, atomic ensembles, and even with micro-mechanical oscillators, with pulsed light. In my talk, I will present our recent theoretical studies towards extending this to continuous-wave light which may provide new perspectives for experiments operating in the regime of strong cooperativity of light-matter interactions. In particular, I will show that this approach can be used to generate deterministically long-distance entanglement of material degrees of freedom, emulate many-body quantum dynamics, and perform analog variational calculations for models of quantum field theories.

Title: Quantum control of ion-photon interactions in an optical cavity

Tracy Northup
University of Innsbruck
Email: tracy.northup@uibk.ac.at

Abstract: By coupling ions to an optical cavity, we can control light–matter interactions at the level of single quanta [1]. I will discuss two contexts within which we are currently investigating quantum control in an ion–cavity QED system.

First, I will consider the transfer of quantum information from ions to photons and vice versa, with the goal of linking distant qubits in a quantum network. A cavity provides a coherent interface with which information can be mapped between an ion and a photon [2]. In recent work, we have shown that encoding a qubit in a superradiant two-ion state enhanced the transfer process by providing protection against spontaneous emission [3]. Looking towards future networks, I will outline how adiabatic passage between two distant atom-cavity systems can mitigate the effects of photon loss in the cavities, thus optimizing the state transfer fidelity for realistic physical systems [4].

Second, I will discuss the nondestructive reconstruction of the cavity photon number, based on the dispersive interaction of the cavity field with a single calcium ion. From Ramsey spectroscopy of the ion, we determine the shift and the broadening of the atomic levels induced by the cavity field. These measurements enable reconstruction of the photon number distribution via a maximum likelihood method [4]. This reconstruction can be seen as the starting point for quantum non-demolition measurement of the optical cavity field and for feedback onto the field.

[1] H. J. Kimble, Nature 453, 1023 (2008)
[2] A. Stute et al., Nat. Photon. 7, 219 (2013)
[3] B. Casabone et al., Phys. Rev. Lett. 114, 023602 (2015)
[4] B. Vogell et al., arXiv:1704.06233
[5] M. Lee et al., in preparation

Title: Kalman Decomposition of Quantum Linear Systems

Guofeng Zhang
The Hong Kong Polytechnic University
Email: guofeng.zhang@polyu.edu.hk

Abstract: Kalman decomposition is a standard result in classical linear systems theory. In this talk, we dis-cuss how to develop Kalman decomposition for quantum linear systems. We show how this structural decomposition can be used to investigate decoherence-free modes, quantum non-demolition variables, and back-action evasion phenomenon in quantum linear systems. Examples from quantum optics and opto-mechanical systems will be used for demonstrations.

Title: Tomography of an Optomechanical Oscillator via Position Measurement

Andrew Doherty
University of Sydney
Email: andrew.doherty@sydney.edu.au

Abstract: We propose an experimental protocol for the quantum state tomography of nonclassical states in optomechanical systems that employs resonant parametric amplification. The procedure overcomes the challenges of weak optomechanical coupling and thermal noise to provide a perfect efficiency homodyne measurement in the limit, which allows for high fidelity tomographic reconstruction of the initial state. Our analysis is based on an analytic description of the measurement (the POVM) that results from integrating the measurement record during the continuous measurement. This analysis fully accounts for the effects of parametric amplification and heating of the oscillator during the measurement. The key technical result is a solution, using a thermofield double approach, of the linear stochastic master equation that may have other applications.

Title: Photonic controlled-phase gates through Rydberg blockade in optical cavities

Etienne Brion
Laboratoire Aimé Cotton
Email: etienne.brion@u-psud.fr

Abstract: We propose a scheme for high-fidelity photonic controlled-phase gates using a Rydberg blockade in an ensemble of atoms in an optical cavity. The gate operation is obtained by first storing a photonic pulse in the ensemble and then scattering a second pulse from the cavity, resulting in a phase change depending on whether the first pulse contained a single photon. We show that the combination of a Rydberg blockade and optical cavities effectively enhances the optical nonlinearity created by the strong Rydberg interaction and makes the gate operation more robust. The resulting gate can be implemented with cavities of moderate finesse, allowing for highly efficient processing of quantum information encoded in photons.

Reference: S. Das, A. Grankin, I. Iakoupov, E. Brion, J. Borregaard, R. Boddeda, I. Usmani, A. Ourjoumtsev, P. Grangier, and A. S. Sørensen, Photonic controlled-phase gates through Rydberg blockade in optical cavities, Phys. Rev. A 93, 040303(R), 2016

Title: Quantum Noise Spectroscopy via Open-Loop Control on Qubit Sensors

Lorenza Viola
Dartmouth College
Email: Lorenza.Viola@dartmouth.edu

Abstract: Accurately estimating the spectral properties of environmental noise in open quantum systems is both a prerequisite for quantitative modeling and prediction, and a key step toward optimizing control performance for fault-tolerant quantum computation and quantum-science applications. I will report theoretical and experimental advances in developing open-loop quantum control methods for spectral estimation using qubit-based sensors. I will first show how, for dephasing environments, dynamical decoupling noise spectroscopy protocols may be extended to both non-Gaussian and multi-qubit settings in principle. I will then describe how Slepian-based control modulation may be leveraged to construct optimally band-limited quantum sensing protocols. In particular, this paves the way to multitaper spectral estimation techniques inspired by classical signal processing, as recently demonstrated in trapped-ion experiments.

Title: Parity-time symmetry and chaos transfer in optomechanics

Jing Zhang
Tsinghua University
Email: jing-zhang@mail.tsinghua.edu.cn

Abstract: This talk is divided into two parts. In the first part, as a natural extension of our previous idea of quantum feedback nonlinearization, we study the nonlinearity-enhancement in parity-time-symmetric optomechanical systems and find its applications in low-threshold phonon laser, highly-sensitive detection of mechanical motion, and control of phonon transport. In the second part, our recent experimental results for optical chaos transfer and chaos-mediated stochastic resonance in optomechanical system will be presented.

Reference:

[1] Z. P. Liu, **J. Zhang*** (co-first author), S. K. Özdemir, B. Peng, H. Jing, X.-Y. Lü, C.-W. Li, L. Yang, F. Nori, and Y.-X. Liu, Metrology with PT-symmetric cavities: enhanced sensitivity near the PT-phase transition, *Phys. Rev. Lett.* 117, 110802 (2016).
[2] F. Monifi, **J. Zhang***(co-first auther), Ş. K. Özdemir*, B. Peng, Y.-X. Liu, F. Bo, F. Nori and L. Yang*, Optomechanically-induced stochastic resonance and chaos transfer between optical fields, *Nature Photonics* 10, 399-405 (2016).
[3] **J. Zhang***, B. Peng, S. K. Ozdemir, Y.-X. Liu, H. Jing, X.-Y. Lü, Y.-L. Liu, L. Yang, and F. Nori, Giant nonlinearity via breaking parity-time symmetry: A route to low-threshold phonon diodes, *Phys. Rev. B* 92, 115407 (2015).
[4] H. Jing, S. K. Özdemir, X.-Y. Lü, **J. Zhang**, L. Yang, and F. Nori, PT-symmetric phonon laser, *Phys. Rev. Lett.* 113, 053604 (2014).
[5] **J. Zhang***, Y.-X. Liu, R.-B. Wu, K. Jacobs, and F. Nori, Quantum feedback: theory, experiments, and applications, *Physics Reports* 679, 1-60 (2017).

Title: Identification and estimation in quantum input-output systems

Madalin Guta
University of Nottingham
Email: Madalin.Guta@nottingham.ac.uk

Abstract: This talk deals with the problem of identifying and estimating dynamical parameters of continuous-time quantum open systems, in the input-output formalism. I will discuss several aspects of this problem:

The first aspect concerns the structure of the space of identifiable parameters for ergodic dynamics, assuming full access to the output state for arbitrarily long times.
I will show that the equivalence classes of undistinguishable parameters are orbits of a Lie group acting on the space of dynamical parameters.

The second aspect concerns the information geometric structure on this space. I will show that the space of identifiable parameters is the base space of a principal bundle given by the action of the group, and carries a Riemannian metric based on the quantum Fisher information of the output. The metric can be computed explicitly in terms of the Markov covariance of certain fluctuation operators, and relate it to the horizontal bundle of the connection.

The third direction concerns the identification of linear input-output systems in the time-dependent and stationary setups.

The fourth direction concerns the design of measurements to achieve the quantum Fisher information.

Title: Echoes of Entanglement: from Quantum Metrology to Scrambling

Monika Schleier-Smith
Stanford University
Email: schleier@stanford.edu

In the quest to approach the fundamental Heisenberg Limit in precision measurements, central challenges are the generation and detection of highly entangled states. I will describe how both of these challenges can be mitigated by "echo spectroscopy," a technique inspired by the Loschmidt echo, a paradigmatic probe of chaos. A key ingredient is to effectively reverse the flow of time in an interacting many-body system by switching the sign of the Hamiltonian. I will describe progress towards engineering spin models with non-local, switchable-sign interactions using cold atoms strongly coupled to light an optical cavity. Intriguingly, similar non-local interactions feature in models for understanding fundamental bounds on chaos and information scrambling derived from the study of black holes, opening prospects for investigating these bounds in the laboratory.

Title: All-optical Computational Circuits in Integrated Photonics

Raymond G Beausoleil
HP Labs
Email: ray.beausoleil@hpe.com

Abstract: We have investigated the potential for scalability of all-optical computational accelerators implemented in photonic integrated circuits, focusing on circuits based on two fundamental physics properties: a) coherence and b) all-optical, resonator-enhanced nonlinearities in bulk materials (such as Kerr, free carrier dispersion and/or thermal nonlinearities). Whereas coherence effects have often impaired the development of scalable implementations of optical computational circuits in free space or fiber based systems, integrated photonics finally offers a platform that can leverage interference effects to allow for scalable computation. As these accelerators process data in the optical domain, they will offer latency, speed, and energy advantages by avoiding or postponing the conversion of optical data to the electronic domain. Potential applications include all-optical routing in switch cores in data centers, preprocessing optical data from IoT sensors at the Edge, and neuromorphic computing. In this context, we found an approach to implement coherent Ising machines in photonic integrated circuits, eventually allowing the mapping of very compute-intensive tasks (such as the infamous NP-hard traveling salesman problem) to optical hardware. Importantly, in realistic circuits, fabrication deviations are larger than the accuracy required to operate coherent computational circuits. Therefore, we have verified in simulation that the resulting phase errors can be compensated using a tuning mechanism that directs external heaters to correct critical phase differences and resonance positions. For the CMOS-compatible a-Si platform, we demonstrated a design approach that allows fabrication of 1000-component all-optical computational circuits in a scalable way, by emphasizing risk mitigation of the non-idealities caused by material imperfections in these advanced, but sometimes still immature material platforms. Research into circuits of this scale was new to the field of photonics: several years ago, circuits would typically contain less than 20 optical components. To enable layout of circuits with ~1000 components, we developed a highly capable photonic layout flow that will help photonic design for interconnects, sensors, and computation.

Title: Control and Readout of Semiconductor Double Quantum Dot Qubits using Circuit Quantum Electrodynamics (QED)

Andreas Wallraff
ETH Zurich, Zurich, Switzerland
Email: andreas.wallraff@phys.ethz.ch

Abstract: Both superconducting quantum bits and semiconductor quantum dots possess excitations in the microwave frequency domain. For this domain we have developed a wide range of novel approaches to create, store, manipulate and detect individual photons using micro-fabricated superconducting quantum electronic circuits. A key asset of this approach are coplanar wave guide resonators in which the field energy of an excitation is distributed over a mode volume much smaller than that of a mirror based resonator. This feature creates sizable electromagnetic fields at the level of individual microwave photons mediating strong electromagnetic interactions. In an approach known as circuit quantum electrodynamics (QED) we probe fundamental quantum optical effects and implement all basic elements required for realizing a quantum information processor. In this talk, I will discuss experiments in which we explore the physics of semiconductor quantum dots in the context of circuit QED. We demonstrate the coherent dipole coupling of double quantum dots to microwave photons [1,2] and detect radiation emitted from the dots in inelastic electron tunneling processes [3]. We also present first measurements of the coherent manipulation of individual double quantum dot charge qubits employing a circuit QED based dispersive readout. This approach provides a basis for the exploration of coherent microwave photon mediated interactions between quantum dot qubits and also for realizing interfaces between different solid state quantum systems in the context of quantum science, technology and engineering.

[1] Dipole coupling of a double quantum dot to a microwave resonator, T. Frey, P. J. Leek, M. Beck, A. Blais, T. Ihn, K. Ensslin, and A. Wallraff, Phys. Rev. Lett. 108, 046807 (2012)
[2] Strong Coupling Cavity QED with Gate-Defined Double Quantum Dots Enabled by a High Impedance Resonator, A. Stockklauser, P. Scarlino, J. V. Koski, S. Gasparinetti, C. K. Andersen, C. Reichl, W. Wegscheider, T. Ihn, K. Ensslin, and A. Wallraff, Phys. Rev. X 7, 011030 (2017)
[3] Microwave Emission from Hybridized States in a Semiconductor Charge Qubit, A. Stockklauser, V. F. Maisi, J. Basset, K. Cujia, C. Reichl, W. Wegscheider, T. Ihn, A. Wallraff, and K. Ensslin, Phys. Rev. Lett. 115, 046802 (2015)

Oscillators for quantum information: the cat code, error correction, and two mode entanglement

Robert Schoelkopf[1] and the Yale circuit QED team

[1]Department of Applied Physics, Yale University, United States

The creation, manipulation, and measurement of non-classical states of light in oscillator modes has been an area of interest in atomic physics and cavity QED for many years. In quantum information, the canonical building blocks are usually qubits realized with two-level atoms or spins, which are then coupled together and arrayed to make more complex, higher-dimensional systems. An oscillator or cavity, however, can store multiple photons and thus serve as more complex building block, or even a logical qubit. Superconducting microwave cavities can also have long coherence times, and offer potential simplifications in the hardware requirements for error correction and other tasks. An example that our team at Yale has been pursuing is the "cat code," [1] which utilizes superpositions of coherent states to create a logical quantum bit that can be efficiently corrected for photon loss. I will present recent experimental progress on using cavities for quantum information, using the coupling to an ancillary superconducting "transmon" qubit. The first experiments [2] demonstrate the redundant encoding of quantum information using multi-photon states. The second set of experiments [3] perform rapid, QND measurements of the photon number parity, which is used to monitor and correct the naturally occurring decoherence, thereby extending the lifetime of quantum information by error correction for the first time. By using a transmon to couple to two cavity modes, we demonstrate the ability to create two-mode cat states [4], or the entanglement of two logical qubits in a Hilbert space of about 150 dimensions. Finally, using techniques from optimal control, we find efficient pulse sequences that can manipulate the encoded information, and we realize [5] high fidelity (>99%) universal gates on the cat code states in a cavity. These experiments show the potential of hardware-efficient approaches to quantum information processing, and open new capabilities for fundamental experiments in quantum optics.

[1] Z. Leghtas et al., Phys. Rev. Lett. **111**, 120501 (2013).
[2] B. Vlastakis, G. Kirchmair, et al., Science **342**, 607 (2013).
[3] N. Ofek, A. Petrenko, et al., arxiv/1602.04768 (2016).
[4] C. Wang, Y. Gao, et al., Science **352**, 1087 (2016).
[5] R. Heeres, P. Reinhold, et al., in preparation (2016).

Control of Quantum Spin Devices, feedback control laws and hidden feedback

Sophie Schirmer, Frank Langbein, Edmond Jonckheere

Networks of interacting spin-1/2 particles form the basis for a wide range of quantum technologies including quantum communication, simulation and computation devices. Optimal control provides methods to steer their dynamics to implement specific quantum operations. It is usually implemented to find optimal time-dependent control fields to implement quantum gates or transformations of quantum states or observables in the context of open-loop quantum control.

We recently proposed an alternative approach of static controls, based on shaping the energy landscape of quantum systems. For coupled spin systems this type of control could be realized in terms of spatially distributed gates that introduce energy level shifts using quasi-static local electric or magnetic fields. Although there are insufficient control degrees of freedom for the system to be completely controllable, many practically interesting operations can be implemented using these controls, including efficient transfer of excitations in spin networks. Furthermore, the resulting controllers combine high fidelity and strong robustness properties under device uncertainties, surpassing traditional limits in classical control. In particular, we observe positive correlations between the logarithmic sensitivity and the control error in many cases, i.e., the highest fidelity controllers are also the most robust. Structured singular value analysis shows the same trend for large structured variation using μ-analysis tools.

One way to understand the surprising robustness of the controllers is in terms of feedback control laws. The energy biases create direct feedback loops. Similar to feedback loops in electronic circuits such as operational amplifiers, this feedback is highly effective and does not require measurements.

I will discuss results on energy landscape control for quantum spin devices with a focus on robustness at high fidelities of the operations and the interpretation of the controllers in terms of feedback control laws.

References:

[1] Emergence of Classicality under decoherence in robust quantum transport, S. Schirmer, E. Jonckheere, S. O'Neil, and F. Langbein, in preparation.

[2] Design of Feedback Control Laws for Information Transfer in Spintronics Networks, S Schirmer, E Jonckheere, F Langbein, arXiv:1607.05294, 2016.

[3] Time optimal information transfer in spintronics networks, FC Langbein, S Schirmer, E Jonckheere, 2015 IEEE 54th Annual Conference on Decision and Control (CDC), 6454-58, 5, 2015.

[4] Structured singular value analysis for spintronics network information transfer control, E Jonckheere, S Schirmer, F Langbein, IEEE Transactions on Automatic Control, DOI: 10.109/TAC.2017.2714623, arXiv:1706.03247, in press, 2017.

[5] Jonckheere-Terpstra test for nonclassical error versus log-sensitivity relationship of quantum spin network controllers, E Jonckheere, S Schirmer, F Langbein, arXiv:1612.02784, 2016.

[6] Information transfer fidelity in spin networks and ring-based quantum routers, E Jonckheere, F Langbein, S Schirmer, Quantum Information Processing 14 (12), 4751-4785.

[7] Multi-fractal Geometry of Finite Networks of Spins, P Bogdan, E Jonckheere, S Schirmer, arXiv:1608.08192, 2016.

Experimental control of two fixed-frequency superconducting qubits coupled to a resonator to create a two-qubit non-adiabatic non-Abelian geometric gate

D. J. Egger, M. Ganzhorn, P. Mueller, A. Fuhrer, S. Filipp
IBM Research – Zurich, 8803 Ruschlikon, Switzerland
Email: DEG@zurich.ibm.com

Fixed-frequency superconducting qubits are controlled with microwave pulses both to generate single-qubit rotations and to activate entangling interactions. A two-qubit entangling gate, needed to form a universal set of logic gates, can be created in many ways. Here we show how to control the $|fg0\rangle \leftrightarrow |gg1\rangle$ and $|gf0\rangle \leftrightarrow |gg1\rangle$ transitions between a resonator and two qubits to create a two-qubit non-adiabatic non-Abelian geometric gate. These microwave-induced transitions have a tunable coupling strength and phase, and serve to transfer an excitation from the second excited state of a qubit to the first excited state of the resonator and vise versa. $|f\rangle$ and $|g\rangle$ respectively denote the second excited state and the ground state of the qubit. $|0\rangle$ and $|1\rangle$ are the first and second resonator states. The gate is activated by simultaneously driving these two transitions. Arbitrary rotations in a two-qubit subspace can be created by changing the amplitudes and phases of the two drives. We also discuss the effects of nearby unwanted transitions such as $|ff0\rangle \leftrightarrow |gf1\rangle$ as well as $|ff0\rangle \leftrightarrow |fg1\rangle$ and what control methods to use to deal with them. These geometric gates are well suited for fixed frequency qubit systems since they work irrespective of the frequencies of the qubits and resonators. This two-qubit gate combined with non-Abelian geometric single qubit rotations completes the necessary toolbox for a quantum computer based on non-Abelian geometric gates.

Open Quantum Systems with arbitrary initial conditions: theory and applications

G. Paz-Silva, M. Hall and H. M. Wiseman
Griffith University, Australia

The theory of Open Quantum Systems is concerned with the prediction and control of the dynamics of a quantum system in the presence of interactions with external degrees of freedom, i.e., the bath. This is generally a highly non-trivial problem, and in its study two assumptions are usually made:

(A1) the state of system and bath is factorizable at time t=0, and

(A2) certain knowledge of the bath, e.g., of the bath correlations, is assumed. This is despite the bath being, by definition, not being fully accessible.

Seeking to replace the assumption character of (A2) by measurable information, recent years have seen the emergence of the so-called Quantum Noise Spectroscopy protocols. These, provided (A1) holds, use the measurable response of a quantum system to its bath and different control scenarios, in order to extract information regarding the bath correlations (with respect to the reduced density matrix of the bath).

In this poster, by introducing a new universal density matrix decomposition we show how assumption (A1) can be naturally removed from current methods to compute the dynamics of a system in the Open Quantum System scenario, such as master equations. Additionally, we show how the notion of Quantum Noise Spectroscopy and more generally Quantum Sensing protocols, can be naturally extended to the case where system and bath are initially correlated. Thus, we show how (A1) and (A2) can in principle be simultaneously discarded from the set of assumptions made in the theory of Open Quantum Systems.

As a key application, we present a Limited Access Quantum System characterization protocol, capable of simultaneously doing both multiqubit state tomography and Hamiltonian learning when fast control and measurements are only available in a subset of them.

Application of adiabatic following to three specific three-sate quantum system

Wei Huang[1], **Bruce W Shore**[2], **Andon Rangelov**[3], **Elica Kyoseva**[4]

[1] *Singapore University of Technology and Design, 8 Somapah Road, 487372 Singapore*

[2] *618 Escondido Cir., Livermore, CA, USA*

[3] *Sofia University, James Bourchier 5 blvd., 1164 Sofia, Bulgaria*

[4] *Institute of Solid State Physics, Bulgarian Academy of Sciences, 1784 Sofia, Bulgari*

Adiabatic time-evolution found in various forms of adiabatic following and adiabatic passage is often advantageous for controlled manipulation of quantum systems due to its insensitivity to deviations in the pulse shapes and timings. In this paper we discuss controlled adiabatic evolution of a three-state quantum system, a natural advance to the widespread use of two-state systems in numerous contemporary applications. We discuss, and illustrate, not only possibilities for population transfer but also for creating, with prescribed relative phase, 50:50 superpositions of two Zeeman sublevels in a letter-vee coupling linkage.

Adiabatic evolution is a widely used quantum control technique that has found numerous applications in different atomic systems, including two-level, three-level, and multi-level quantum systems. One of the main advantages of adiabatic following is its robustness to varying Rabi frequencies and/or detunings. Previous work in three-level configurations showed only specific transfer patterns such as starting from the excited state. In this work, we overcome this constriction and show how to implement adiabatic following in a general three-state quantum system. The evolution of a three-state systems is described by the Schrödinger equation (in RWA),

$$i\frac{d}{dz}C(t) = \begin{bmatrix} -\Delta(t) & \frac{1}{2}\widetilde{\Omega}(t) & 0 \\ \frac{1}{2}\widetilde{\Omega}^*(t) & 0 & \frac{1}{2}\widetilde{\Omega}(t) \\ 0 & \frac{1}{2}\widetilde{\Omega}^*(t) & \Delta(t) \end{bmatrix} C(t), \tag{1}$$

where $C(t)$ is the amplitude vector of states 1, 2 and 3, and $\widetilde{\Omega}(t) = \Omega(t)e^{i\phi}$ is the Rabi frequency with ϕ is a phase that can be externally controlled by a laser pulse.

When the initially populated state is state 1 or 3, and we have a large detuning, then the state of the system coincides with an adiabatic state $|a_-\rangle$ or $|a_+\rangle$. Then, by increasing adiabatically the Rabi frequency, the systems evolves into state $|a_\pm\rangle$. This translated to 50% population in state 2 and 25% in both states 1 and 3. We present this scenario in Fig. 2 (a) where the coupling parameters are $\Omega(t) = \Omega_0\exp[-(t-\tau)^2/T^2]$ and $\Delta(t) = \Delta_0\exp[-(t+\tau)^2/T^2]$.

For the same initial state, but linear with time detuning the evolution of the system is different and we show it in Fig. 2 (b). The coupling parameters are $\Omega(t) = \Omega_0\exp(-t^2/T^2)$ and $\Delta(t) = \delta_0 + \Delta_0 t$. For this evolution, at $t \to 0+$, we have a crossing of the adiabatic states. Finally, due to that complete population transfer is obtained.

Finally, starting at state 2, we start in state $|a_0\rangle$ and then adiabatically transfer to lager Rabi frequency. We show the final state superposition in Fig 2 (c). with coupling strength and detuning the same for Fig 2 (a). More details of this work can be found in ref.[1].

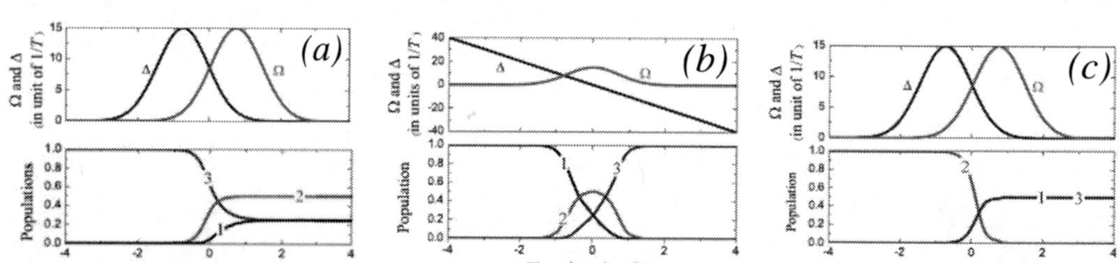

Figure 1: The coupling strength $\Omega(t)$ and detuning $\Delta(t)$, and the corresponding population evolution.

References [1] W. Huang, B. W. Shore, A. Rangelov and E. Kyoseva, Optics Communications **382**, 196-200 (2017).

Towards the Fundamental Quantum Limit of Linear Measurements of Classical Signals

Haixing Miao*, Rana X Adhikari, Yiqiu Ma, Belinda Pang, Yanbei Chen

Abstract:

The quantum Cramér-Rao bound (QCRB) sets a fundamental limit for the measurement of classical signals with detectors operating in the quantum regime. Using linear-response theory and the Heisenberg uncertainty relation, we derive a general condition for achieving such a fundamental limit. When applied to classical displacement measurements with a test mass, this condition leads to an explicit connection between the QCRB and the Standard Quantum Limit which arises from a tradeoff between the measurement imprecision and quantum backaction; the QCRB can be viewed as an outcome of a q uantum non-demolition measurement with the backaction evaded. Additionally, we show that the test mass is more a resource for improving measurement sensitivity than a victim of the quantum backaction, which suggests a new approach to enhancing the sensitivity of a broad class of sensors. We illustrate these points with laser interferometric gravitational wave detectors.

Reference: arXiv:1608.00766

*Email: haixing@star.sr.bham.ac.uk

Thermodynamics along individual trajectories of a quantum bit

M. Naghiloo, D. Tan, P. M. Harrington, J. J. Alonso, E. Lutz, A. Romito, K. W. Murch

The ability to continuously monitor quantum systems opens a new horizon to explore quantum mechanics in small quantum systems at the level of single trajectories. Here we use resonance fluorescence trajectories to study quantum dynamics and quantum thermodynamics in the context of quantum measurement. We use a near-quantum-limited detector to experimentally track individual quantum trajectories of a driven qubit formed by the hybridization of a waveguide cavity and a transmon circuit. For a single trajectory, we separately identify energy changes of the qubit as heat and work, and verify the first law of thermodynamics for an open quantum system. We further employ a novel quantum feedback loop to compensate for the exchanged heat and effectively isolate the qubit. By verifying the Jarzynski equality for the distribution of applied work, we demonstrate the validity of the second law of thermodynamics. Our results establish thermodynamics along individual quantum trajectories.

Keywords: Quantum thermodynamics, Quantum trajectory, resonance florescence trajectory, Jarzynski equality.

Classical optical logic at the single photon level with diamond nanophotonic networks

Yaser Silani, Forrest Hubert, and Victor Acosta, University of New Mexico

What are the fundamental limits of optical computing, and is there a scalable platform that can be used to test them? In this project we seek to probe the fundamental limits of optical logic at both the nanometer and single-photon scales. Our platform for probing these limits uses all-optical circuits fabricated from color-center-doped diamond films. This approach differs from previous failed attempts by using integrated nanophotonic networks to perform logic with the lowest power consumption allowed by quantum mechanics-the single photon level. Logical operations will be performed by nonlinear optical addressing [1,2] of Silicon Vacancy (SiV) color centers in diamond that are coupled to nanophotonic cavities. SiVs in diamond are promising candidates for quantum nonlinear optics because of their narrow zero-phonon-lines in the near infrared, weak phonon coupling, and resilience against electric-field induced broadening.

Sequential logic will be enabled by embedding SiV-doped cavities into interacting networks of optical waveguides, beamsplitters, and other on-chip photonic elements. We are now routinely fabricating the basic building blocks of nanophotonic networks in the cleanrooms at UNM and CINT. We have assembled a cryogenic confocal microscope with high spectral resolution (<1 GHz) for device spectroscopy and characterization. Recently we used this platform to observe the narrow SiV fine structure in thin diamond films at cryogenic temperatures. In our next steps, we will begin proof-of-principle all-optical switching experiments in nanophotonic devices.

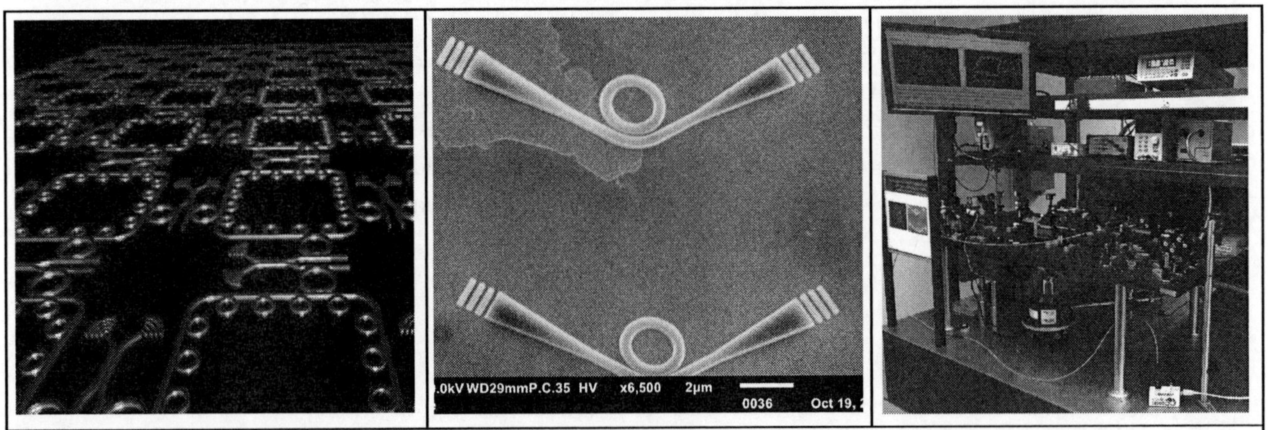

Quantum limits of classical optical logic. (left) A scalable nanophotonic network to test the fundamental quantum limits of classical optical computing. (middle) SEM of diamond photonic resonators coupled to a waveguide with grating couplers. (right) Part of the cryogenic confocal microscope at UNM.

[1] **VM Acosta**, K Jensen, C Santori, D Budker, RG Beausoleil, PRL (2013).
[2] A Sipahigil et al., Science (2016).

An experimental system for quantum simulation using $^{171}Yb^{+}$ ions

A. R. Milne[1*] , C. Hempel[1] , M. J. Biercuk[1]
[1]*ARC Centre for Engineered Quantum Systems, School of Physics,*
The University of Sydney, Sydney, New South Wales 2006, Australia.
**Email: amil1264@uni.sydney.edu.au*

We present an experimental system designed to perform quantum simulation experiments using trapped $^{171}Yb^{+}$ ions in a Paul trap; two classes of experiments will be discussed. The first involves phase-modulated pulse sequences for minimising residual qubit-oscillator entanglement at the conclusion of two-qubit entangling gates [1]. Secondly, we present a protocol for 'programmable' quantum simulation of Ising-type Hamiltonians, specifically, one-dimensional spin chains with translation-invariant symmetrically coupled two-body terms (i.e. XX-, YY-, ZZ- type couplings) [2]. Beginning with native long-range pairwise interactions, we show that short sequences of Pauli operators applied to individual spins can effectively generate any other Hamiltonian in this family, with the required sequence length scaling polynomially with the number of spins.

In our system, long-range pairwise interactions between spins are engineered via coupling to a shared mode of motion. This is done using a 355 nm pulsed laser system to perform two-photon Raman-mediated entangling gates. The geometry of the 355 nm beam paths allows for both global and single ion addressing. For global operations, the entire ion chain is illuminated by two counter-propagating beams delivered via UV optical fibers through holes in the trap endcaps. Single qubit operations are achieved via diffraction-limited focusing of an additional 355 nm beam path through a custom objective lens (NA 0.56) suspended above the chamber. This lens also enables diffraction-limited imaging of the ions at 369 nm and is used for high fidelity state detection. In addition to this optical gate capability, the system incorporates an in-vacuum loop antenna resonant with the 12.6 GHz hyperfine qubit transition.

References:

[1] T. J. Green and M. J. Biercuk, Phys. Rev. Lett. 114, 120502 (2015).

[2] D. Hayes, S. T. Flammia, and M. J. Biercuk, New Journal of Physics 16, 083027 (2014).

In situ upgrade of quantum simulators to universal computers

Ben Dive[1, 2], Alexander Pitchford[2], Florian Mintert[1], and Daniel Burgarth[2]

[1] *Controlled Quantum Dynamics, Imperial College London, UK, SW7 2AZ*
[2] *Institute of Mathematics, Physics and Computer Science, Aberystwyth University, UK, SY23 3BZ*

May 9, 2017

Recent and ongoing work on building large quantum systems is leading to simulators that are able to model physical phenomena, allowing questions about the underlying science to be answered. The presence of interactions between the qubits that make up the simulator leads to dynamics that, by varying control parameters in the system Hamiltonian, can replicate the quantum behaviour of systems of interest. This, however, is less general than a quantum computer which is able is to perform a universal set of logic gates on the qubits and thereby arbitrary algorithms.

These gates are normally possible to realise on the simulator theoretically, provided that there are some control parameters can be varied in time. The difficulty lies in finding the right time-dependency, which relies almost exclusively on numerical methods, especially when physical constraints on the control fields are taken into account. These require a very precise knowledge of the parameters of a system, a daunting task for a machine with a huge number of degrees of freedom, and the numerical methods themselves are intractable on a classical computer if the quantum machine is large enough to be useful.

We circumvent both these problems by showing how well-known existing numerical methods can be translated to run *in situ* on the quantum simulator itself. The quantum simulator is used to evaluate how close a set of control parameters lead the system to the desired evolution, information which is then fed to a classical computer which decides what control parameters to try next. This is a bootstrapping scheme for learning a universal set of gates which needs classical computational and experimental resources that scale favourably. Analytical results on measuring gate fidelities approximately, combined with extensive numerical simulations, suggest that the overall cost of this method is polynomial in the number of qubits. This makes the proposal a feasible approach to turn state-of-the-art and near-future quantum simulators with tens of qubits into universal quantum computers.

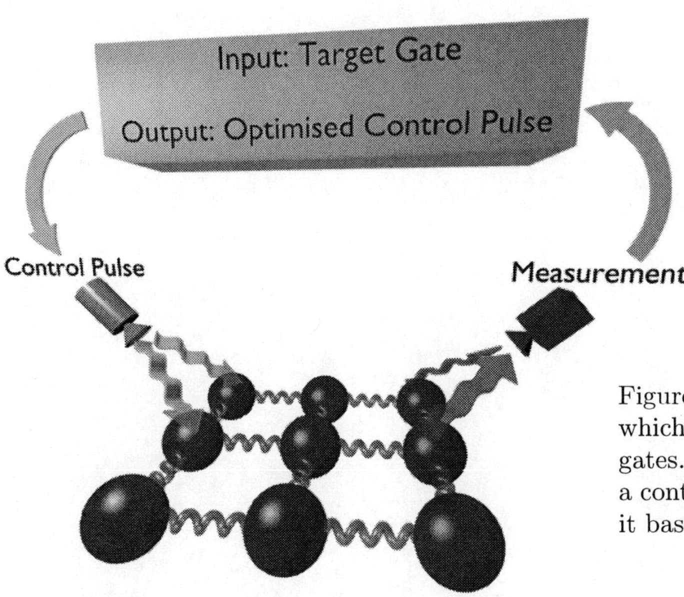

Figure 1: A classical computer finds a control pulse which enables a quantum simulator to perform logic gates. It does this in an iterative process by applying a control pulse to the simulator and then improving it based on the result of measurements.

Entanglement Preservation using Feedback Control in a Heisenberg Spin Chain

Jin Wang
Department of Physics, The University of Michigan, Dearborn, MI 48120

This work studies the ability of feedback control to reduce the effect of decoherence and preserve entanglement in an interacting Heisenberg chain model in the presence of an external magnetic field. The system reaches an improved steady state entanglement when feedback is present. The influence of the strength of the external B field is found for optimized steady state entanglement. The time-dependent entanglement evolution of the system is also studied. One potential application of this scheme is to raise the maximum operating temperature of spin entangled system.

Cooling of a laser levitated nanoparticle in the shot noise dominant regime

Changchun Zhong[1]
[1]Department of Physics and Astronomy, Purdue University, West Lafayette IN, 47907 USA

Quantum and classical versions feedback cooling of an optically trapped anisotropic nanoparticle in the laser shot noise dominant regime are numerically investigated. The quantum calculations using the theory of continuous quantum measurement quantitatively match the classical simulation. In the laser levitated nano system, we also show that the related dynamical parameters, such as the oscillating frequency and shot noise heating rate, depend on the shape of the trapped particle. For two different feedback cooling schemes, we also get results on the lowest occupation number that can be achieved as a function of the heating rate and the amount of measurement uncertainty in the position. As the interests on ground state cooling of anisotropic nanoparticles increase, our study can serve as a framework for thinking about how those parameters affect the heating and the feedback cooling of levitated nanoparticles.

[1] C. Zhong and F. Robicheaux, Phys. Rev. A, vol. 94, p. 052109, Nov 2016. "Decoherence of rotational degrees of freedom" .

[2] C. Zhong and F. Robicheaux, arXiv:1701.04477 "Shot noise dominant regime of a nanoparticle in a laser beam" .

[3] C. Zhong and F. Robicheaux , In preparation "Feedback cooling of laser levitated nanoparticles through continuous quantum measurement".

Experimental quantum Hamiltonian learning

Jianwei Wang, Stefano Paesani, Raffaele Santagati, Sebastian Knauer, Antonio A. Gentile, Nathan Wiebe, Maurangelo Petruzzella, Jeremy L. O'Brien, John G. Rarity, Anthony Laing and Mark G. Thompson

The efficient characterization of quantum systems, the verification of the operations of quantum devices and the validation of underpinning physical models are central challenges for quantum technologies and fundamental physics. The computational cost of such studies could be improved by machine learning enhanced by quantum simulators. Here we interface two different quantum systems through a classical channel—a silicon-photonics quantum simulator and an electron spin in a diamond nitrogen–vacancy centre—and use the former to learn the Hamiltonian of the latter via Bayesian inference. We learn the salient Hamiltonian parameter with an uncertainty of approximately 10^{−5}. Furthermore, an observed saturation in the learning algorithm suggests deficiencies in the underlying Hamiltonian model, which we exploit to further improve the model. We implement an interactive version of the protocol and experimentally show its ability to characterize the operation of the quantum photonic device.

Application of optimal band-limited control protocols to quantum noise sensing

Poster abstract for PRACQSYS 2017

Virginia Frey, University of Sydney, 14/05/2017

Nanoscale sensing techniques are in increasing demand across medical, metrological and industrial applications. In the context of quantum coherent devices, sensing protocols are deployed to exploit the system's sensitivity to the environment as a means to augment sensor performance [1]. However, common sensing protocols for qubit-based sensors, as implemented through control operations that manipulate their response in frequency space, suffer from out-of-band spectral leakage which complicates the interpretation of a sensor's signal [2].

We tackle this challenge and demonstrate [3] a novel type of sensing protocols based on the provably optimal band-limited Slepian functions [4]. Experiments with trapped ion qubits deploying this method reveal orders of magnitude improvement over conventional sensing protocols. Combining the narrowband control protocols with concepts from RF engineering to tune the frequency sensitivity band of our qubits, we demonstrate how complex coherent amplitude noise spectra can be reconstructed using classical post-processing techniques like multitaper and Bayesian spectrum estimation. Contributions to qubit infidelity due to noise in different axes are identified using tomographic measurements. Lastly, we apply these techniques to measure the intrinsic noise in our qubit-drive synthesis chain with sensitivities down to 0.001 dB.

References

[1] L. P. McGuinness, L. T. Hall, A. Stacey, D. A. Simpson, C. D. Hill, J. H. Cole, K. Ganesan, B. C. Gibson, S. Prawer, P. Mulvaney, F. Jelezko, J. Wrachtrup, R. E. Scholten, and L. C. L. Hollenberg, "Ambient nanoscale sensing with single spins using quantum decoherence," *New J. Phys.*, vol. 15, no. 7, p. 073042, 2013.

[2] M. Loretz, J. M. Boss, T. Rosskopf, H. J. Mamin, D. Rugar, and C. L. Degen, "Spurious harmonic response of multipulse quantum sensing sequences," *Phys. Rev. X*, vol. 5, p. 021009, Apr 2015.

[3] V. M. Frey, S. Mavadia, L. M. Norris, W. de Ferranti, D. Lucarelli, V. Lorenza, and M. J. Biercuk, "Application of optimal band-limited control protocols to quantum noise sensing," *arXiv preprint arXiv:1704.02050*, 2017.

[4] D. Slepian and H. O. Pollak, "Prolate spheroidal wave functions, fourier analysis, and uncertainty -I," *Bell Labs Technical Journal*, vol. 40, no. 1, pp. 43–63, 1961.

Ultimate precision limit for quantum parameter estimation

Haidong Yuan
Chinese University of Hong Kong

ABSTRACT

Measurement and estimation of parameters are essential for science and engineering, where the main quest is to find out the highest achievable precision with given resources and design schemes to attain it. With recent development of technology, it is now possible to design measurement protocols utilizing quantum mechanical effects, such as entanglement, to attain far better precision than classical schemes. This has found wide applications in quantum phase estimation, quantum imaging, atomic clock synchronization, etc, and created a high demand for better understanding of measurement protocols based on quantum mechanical effects. I will present a general framework for quantum metrology which relates the ultimate precision limit to the underlying quantum dynamics. This framework provides efficient methods for computing the ultimate precision limit and optimal schemes to attain it. It also provides an analytical formula of the precision limit with arbitrary pure probe states, which spares the need of optimization required in previous studies. It is also shown that with noiseless dynamics a universal time scaling emerges as a fundamental property under the optimal scheme for quantum parameter estimation, this restores an intuition that has been recently questioned in the field, that time is always a valuable resource.

References:
Haidong Yuan, Physical Review Letters 117, 160801 (2016).
Haidong Yuan, Chi-Hang Fred Fung, npj: Quantum Information, 3:14, (2017).

Measuring and Suppressing Error Correlations in Quantum Circuits

C. Edmunds,[1,2] C. Hempel,[1,2] S. Mavadia,[1,2] H. Ball,[1] T. M. Stace,[3] and M. J. Biercuk[†1,2]

[1]*ARC Centre for Engineered Quantum Systems, School of Physics, The University of Sydney, NSW Australia*
[2]*National Measurement Institute, West Lindfield NSW 2070 Australia*
[3]*ARC Centre for Engineered Quantum Systems, School of Physics and Mathematics, The University of Queensland, St Lucia, QLD Australia*
(Dated: Sunday 23rd April, 2017 at 09:05)

Quantum error correction and the theory of fault-tolerance provide a path to large-scale quantum computing, but are challenged by the assumption of errors that are stochastic throughout a quantum circuit. The presence of temporally correlated errors between sequential logic gates violates this requirement, but is a realistic element of laboratory environments where low frequency, strongly correlated noise, such as flicker frequency noise $\propto 1/f$, is commonly present. Consequently, to facilitate the implementation of QEC it is necessary to identify and suppress such errors. We provide an analytic framework that allows the identification of correlated errors in randomly compiled quantum circuits using only projective measurements at the end of the circuit. Using a single trapped $171Yb^+$ ion, we experimentally identify signatures of error correlations in the presence of engineered noise with tuneable temporal correlations. Finally, we demonstrate the suppression of error correlations by replacing standard logic gates with dynamically protected gates, and quantitatively extract a $\sim$16 reduction in the correlated error component compared to standard gates.

GaP-on-Diamond Integrated Photonics Platform for NV-Center-Based Quantum Computation

Srivatsa Chakravarthi, Emma Schmidgall, Alan Logan, Ian R. Christen, Michael Gould and Kai-Mei C. Fu

Optically accessible solid-state defects are heralded as a promising platform for distributed quantum computation. The NV center in diamond is particularly attractive due to its long electron spin-coherence time and access to multiple nuclear spins which could be utilized for topologically protected cluster-state generation [1]. However, a complete platform to efficiently entangle multiple NV centers remains elusive. Here we present recent results in a GaP-on-diamond photonic platform designed for efficient NV-NV entanglement. With a larger refractive index than diamond (3.3 vs. 2.4), GaP photonics allow efficient NV photon collection and routing [2]. In-plane electrodes allow Stark tuning to tune and stabilize the NV optical emission [3]. The ultra-smooth GaP surface profile allows integration of superconducting detectors for efficient on-chip entanglement generation. We demonstrate the fabrication and operation of these components individually and provide a clear roadmap for their integration.

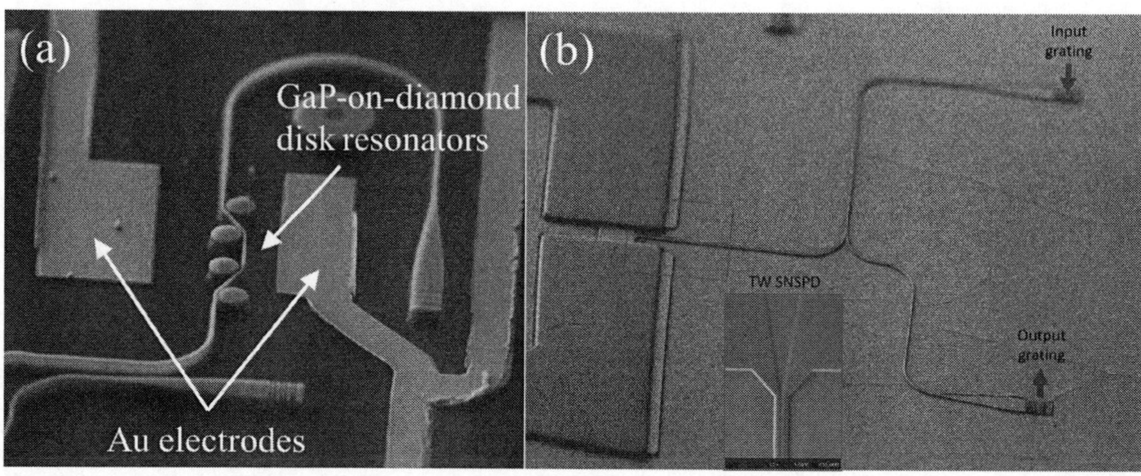

[1] Y. Li and S. C. Benjamin, "High threshold distributed quantum computing with three-qubit nodes," *New J. Phys.*, vol. 14, no. 9, p. 93008, Sep. 2012.

[2] M. Gould, E. R. Schmidgall, S. Dadgostar, F. Hatami, and K.-M. C. Fu, "Efficient Extraction of Zero-Phonon-Line Photons from Single Nitrogen-Vacancy Centers in an Integrated GaP-on-Diamond Platform," *Phys. Rev. Appl.*, vol. 6, no. 1, p. 11001, Jul. 2016.

[3] V. M. Acosta, C. Santori, A. Faraon, Z. Huang, K.-M. C. Fu, A. Stacey, D. A. Simpson, K. Ganesan, S. Tomljenovic-Hanic, A. D. Greentree, S. Prawer, and R. G. Beausoleil, "Dynamic Stabilization of the Optical Resonances of Single Nitrogen-Vacancy Centers in Diamond," *Phys. Rev. Lett.*, vol. 108, no. 20, p. 206401, May 2012.

Title: Measurement of Correlations in a Symmetric Many-Body Quantum State via Continuous Measurement

Abstract: Continuous measurement of an ensemble of spins in the presence of external control has been proven to be a fast and robust way to reconstruct the one-body reduced density matrix of a quantum state [1-5]. We desire to expand this protocol to reconstruct the correlations in a symmetric many-body state of multiple qubits. To accomplish this the many-body system should be probed collectively, weakly enough not to erase the initial conditions over the duration of the measurement, but strongly enough to map the information about the initial state onto the measurement record. To recover the one-body state, it suffices to employ local unitary rotations on the collective state. Moreover, for a sufficient number of spins, one can obtain high fidelity in the reconstruction while measuring for a time that probe shot noise dominates over the spin projection noise. In this case measurement backaction is negligible. By contrast, the fact that the stochastic evolution of the state depends on the projection noise implies that the information about correlations are in principle in the conditional evolution. Deducing the correlation requires us to resolve this over shot noise of the probe, in which case measurement backaction is intrinsic to the protocol. We explore the information-gain/disturbance tradeoff and the requirements for reconstruction of correlations. Thereby we address the question, how much information about the many-body state can one extract with one copy of the state?

References:

[1] Andrew Silberfarb, Poul S. Jessen, and Ivan H. Deutsch, Phys. Rev.Lett. 95, 030402 (2005)
[2] Greg A. Smith, Andrew Silberfarb, Ivan H. Deutsch, and Poul S. Jessen Phys. Rev. Lett. 97, 180403 (2006)
[3] Carlos A Riofrío, Poul S Jessen and Ivan H Deutsch, Journal of Physics B: Atomic, Molecular and Optical Physics, 44, 15 (2011)
[4] A. Smith, C. A. Riofrío, B. E. Anderson, H. Sosa-Martinez, I. H. Deutsch, and P. S. Jessen, Phys. Rev. A 87, 030102(R) (2013)
[5] Robert L. Cook, Carlos A. Riofrío, and Ivan H. Deutsch, Phys. Rev. A 90, 032113 (2014)

Towards generic adiabatic elimination for bipartite open quantum systems

R. Azouit *[1], F. Chittaro[2], A. Sarlette[3], and P. Rouchon[1]

[1]Centre Automatique et Systèmes, Mines-ParisTech, PSL Research University. 60 Bd Saint-Michel, 75006 Paris, France.
[2]Aix Marseille Université, CNRS, ENSAM, LSIS UMR 7296, 13397 Marseille, France, and Université de Toulon, CNRS, LSIS UMR 7296, 83957 La Garde, France
[3]INRIA Paris, 2 rue Simone Iff, 75012 Paris, France; and Ghent University, Belgium.

While adiabatic elimination is standard for closed quantum systems using regular perturbation theory [11], the case of open quantum systems is more complicated. Several particular examples have been treated (see e.g. [1], [4], [10], [6]). A generalisation of the Schrieffer-Wolff formalism has been developed in [9] by treating the Lindblad equation as a standard linear system. A model reduction method for quantum stochastic equations was proposed in [3] when the speed of the fast system goes to infinity, making difficult the characterisation of the order of the approximation.

We present a original approach, based on [2] and related to [12], for adiabatic elimination which provides formulas with improved direct applicability, control on the order of approximation, and explicit expression of how the reduced system is embedded in the full Hilbert space. We focus on bipartite quantum systems. One is highly dissipative i.e. on a fast time-scale while the other one is on a slow time-scale. We investigate the effect of a weak coupling by treating the slow dynamics as a perturbation of the fast one. Using geometric singular perturbation theory [8] and center manifold theory [5], we perform an asymptotic expansion in power of the small parameter describing the different time-scale allowing to readily select the order of the approximation. This approach ensure a Lindblad master equation for the reduced dynamics as well as a Kraus map [7] linking the complete and the reduced model. These two constraints ensure the interpretation in term of density operator of the slow dynamics. We present explicit formulas to compute the reduced dynamics up to second order. These formulas already allow to compute more precise reduced dynamics on relevant example as well as give general structural results on the reduced dynamics.

References

[1] D. J. Atkins, H. M. Wiseman, and P. Warszawski. Approximate master equations for atom optics. *Phys. Rev. A*, 67:023802, Feb 2003.

[2] R. Azouit, F. Chittaro, A. Sarlette and P. Rouchon. Towards generic adiabatic elimination for bipartite open quantum systems. *Submitted, see arXiv:1704.00785.*

[3] L. Bouten and A. Silberfarb. Adiabatic Elimination in Quantum Stochastic Models. In *Commun. Math. Phys.*, 283: 491. doi:10.1007/s00220-008-0513-6, 2008.

[4] E Brion, L H Pedersen, and K Mølmer. Adiabatic elimination in a lambda system. *Journal of Physics A: Mathematical and Theoretical*, 40(5):1033, 2007.

[5] J. Carr. *Application of Center Manifold Theory.* Springer, 1981.

*remi.azouit@ens-cachan.fr

[6] O. Cernotik, D.V. Vasilyev, and K. Hammerer. Adiabatic elimination of gaussian subsystems from quantum dynamics under continuous measurement. *Phys. Rev. A*, 92(1):012124–, July 2015.

[7] M.D. Choi. Completely positive linear maps on complex matrices. *Linear algebra and its applications*, 10:285–290, 1975.

[8] N. Fenichel. Geometric singular perturbation theory for ordinary differential equations. *J. Diff. Equations*, 31:53–98, 1979.

[9] E. M. Kessler. Generalized Schrieffer-Wolff formalism for dissipative systems. *Phys. Rev. A*, 86(1):012126–, July 2012.

[10] M. Mirrahimi and P. Rouchon. Singular perturbations and Lindblad-Kossakowski differential equations. *IEEE Trans. Automatic Control*, 54(6):1325–1329, 2009.

[11] J.J. Sakurai and J. Napolitano. *Modern quantum mechanics*. Addison-Wesley, 2011.

[12] P. Zanardi and L. Campos Venuti. Coherent Quantum Dynamics in Steady-State Manifolds of Strongly Dissipative Systems. *Phys. Rev. Lett.*, 113(24):240406, 2014.

Abstract of the Contributed Poster for PRACQSYS 2017

Dimension estimation assisted by a single quantum probe

Akira Sone and Paola Cappellaro

Research Laboratory of Electronics and Department of Nuclear Science and Engineering,
Massachusetts Institute of Technology,
77th Massachusetts Avenue, Cambridge, MA, 02139

A quantum system's dimension is an important resource to determine the performance of the system in applications to, e.g., quantum computation, quantum sensing or simulation. However, measuring the Hilbert space dimension is a demanding task. Although many dimension witnesses have been proposed, these protocols can only give a lower bound on the dimension by measuring correlations. This poses a severe limitation to practical implementations, particularly for many-body qubit systems with limited accessibility, as one cannot measure all the correlation observables.

Here, we propose a practical strategy for dimension estimation, which relies on the indirect interaction of the target system with a fully accessible, single quantum probe. We combine realization theory with a construction based on recursive families of graphs to demonstrate that the system dimension can be exactly estimated from the model order of the system. The protocol works under the assumption that the interaction model is given and as long as the all the qubits become correlated with the quantum probe during the time evolution. We further perform a robustness analysis in the presence of background noise to evaluate what noise levels still allow estimating the exact system dimension.

Reference:

[1] Akira Sone and Paola Cappellaro, arXiv:1702.03280

Optimal detection of precessing spin ensembles

Ricardo Jiménez-Martínez[1], Vito Giovanni Lucivero [1], Aleksandra Dimic [2],
Charikleia Troullinou [1], Jia Kong [1], Janek Kolodynski[1], and Morgan W. Mitchell [1,2]
[1]*ICFO-Institut de Ciencies Fotoniques, The Barcelona Institute of Science and Technology, 08860 Castelldefels (Barcelona), Spain*
[2] *Faculty of Physics, University of Belgrade, Studentski Trg 12-16, 11000 Belgrade, Serbia*
[3] *ICREA – Institució Catalana de Recerca i Estudis Avançats, 08015 Barcelona, Spain*

Precise knowledge of the spin orientation of atomic ensembles is demanded in a wide range of scenarios, from further progress in best-in-class atomic clocks and magnetometers to the preparation of non-classical states of matter and memories. Implementing quantum-limited detection schemes is challenging, however. Shot-noise in the detectors can mask the motion of the spins limiting the precision of estimations. Here I describe our recent work and results aimed at overcoming this challenge in alkali-based magnetometers. I will describe versatile instrumentation to optically prepare and detect alkali vapors using coherent and polarization-squeezed light [1] and optimal estimation techniques [2]. I will also report our progress in the development and verification of these techniques in the time domain as required for real-time spin state estimation in magnetic field sensing and control applications. The prospects of integrating these techniques with recently developed microfabaricated spin polarized sensors [3, 4] will be discussd as well.

[1] V. G. Lucivero, et al., "Squeezed-light spin noise spectroscopy", Phys. Rev. A **93**, 053802 (2016).
[2] V. G. Lucivero, et al., "Sensitivity, quantum limits, and quantum enhancement of noise spectroscopies", Phys. Rev. A **95**, 041803 (2017).
[3] R. Jimenez-Martinez, et al., "Optical hyperpolarization and NMR detection of 129Xe on a microfluidic chip", Nature. Comm. **5**, 3908 (2014).
[4] D. Kennedy, et al. "An optimized microfabricated platform for the optical generation and detection of hyperpolarized 129Xe." Sci. Reports 7 (2017).

Controlling quantum information in open quantum systems

Victor V. Albert,[1] Chao Shen,[1] Stefan Krastanov,[1] Kyungjoo Noh,[1] Barry Bradlyn,[2] Martin Fraas,[3] R. J. Schoelkopf,[1] Mazyar Mirrahimi,[1,4] S. M. Girvin,[1] Michel H. Devoret,[1] and Liang Jiang[1]

[1] *Yale Quantum Institute and Departments of Applied Physics and Physics, Yale University, New Haven, Connecticut 06511, USA*

[2] *Princeton Center for Theoretical Science, Princeton University, Princeton, New Jersey 08540, USA*

[3] *Instituut voor Theoretische Fysica, KU Leuven, Leuven 3000, Belgium*

[4] *QUANTIC project-team, INRIA of Paris, Paris 75012, France*

(Dated: November 30, 2016)

Lindbladians, one of the simplest extensions of Hamiltonian-based quantum mechanics, are used to describe decay and decoherence of a quantum system induced by the system's environment. While traditionally viewed as detrimental to fragile quantum properties, a tunable environment offers the ability to drive the system toward steady-state subspaces, which can be used to store, protect, and process quantum information. A prominent example of a steady-state subspace is a cat code — a subspace spanned by superpositions of well-separated coherent states and protected from certain errors. We report on two schemes designed to process quantum information in cat codes using Hamiltonian-based [1] and holonomic [2] control. Both schemes have been extended to general steady-state subspaces [3]. While such control schemes can be used for (unitary) gates, they are not sufficient to perform (nonunitary) error correction. We discuss two ways to implement more general nonunitary maps — quantum channels — in order to drive quantum information back into a steady-state subspace after an error. The first is a recipe to embed any channel into the infinite-time evolution generated by a Lindbladian [3]. The second is an efficient implementation for state-of-the-art circuit QED experiments based on adaptive control [4].

---

[1] M. Mirrahimi, Z. Leghtas, V. V. Albert, S. Touzard, R. J. Schoelkopf, L. Jiang, and M. H. Devoret, New J. Phys. **16**, 045014 (2014).

[2] V. V. Albert, C. Shu, S. Krastanov, C. Shen, R.-B. Liu, Z.-B. Yang, R. J. Schoelkopf, M. Mirrahimi, M. H. Devoret, and L. Jiang, Phys. Rev. Lett. **116**, 140502 (2016).

[3] V. V. Albert, B. Bradlyn, M. Fraas, and L. Jiang, Phys. Rev. X **6**, 041031 (2016).

[4] C. Shen, K. Noh, V. V. Albert, S. Krastanov, M. H. Devoret, R. J. Schoelkopf, S. M. Girvin, and L. Jiang, arXiv:1611.03463.

Kalman filtering for predictive control of qubits under dephasing

R. Gupta[1], C. Granade[1], M. J. Biercuk[1,*]

[1] *ARC Centre for Engineered Quantum Systems, School of Physics, The University of Sydney, Sydney, New South Wales 2006, Australia.*
* *Email:* rgup9526@uni.sydney.edu.au

We use Kalman based techniques [1] for predictive control of qubits interacting with a noise field. Our data represent a fluctuating qubit phase measured via a sequence of Ramsey experiments [2]. Learning algorithms can use noise correlations encoded in our data to enable prediction of the stochastic noise forward in time [3]. A key feature of this Kalman implementation is the use of a 'basis of oscillators' to mimic stochastic evolution in the absence of dynamic model. Recursive state estimation allows a calculation of the instantaneous amplitude and the phase for each Kalman basis oscillator enabling true power spectral densities (PSDs) estimation and time domain predictions in a single run.

We test filter performance against imperfect learning environments, measurement noise and aliasing using randomly generated data from complex PSDs. Without a priori knowledge of the true noise field, we optimise filter design parameters (process and measurement noise covariance strengths) using training data [4], [5]. Our results show that Kalman predictions outperform predicting the mean of the true stochastic process in moderately noisy, imperfect learning environments. Further, filter computations appear to possess a distinct signature whenever failure occurs due to aliasing. Extensions to binary projective measurements and data containing non commuting observables will be discussed.

REFERENCES:

[1] Liska, Jindrich, and Eduard Janecek. "Time-Frequency Representation of Signals Using Kalman Filter." *Robotics Automation and Control*. InTech, 2008.
[2] Soare, A., et al. "Experimental bath engineering for quantitative studies of quantum control." *Physical Review A* 89.4 (2014): 042329.
[3] Mavadia, Sandeep, et al. "Prediction and real-time compensation of qubit decoherence via machine learning." *Nature Communications* 8 (2017).
[4] Abbeel, Pieter, et al. "Discriminative Training of Kalman Filters." *Robotics: Science and systems*. Vol. 2. 2005.
[5] Stenberg, Markku PV, Oliver Köhn, and Frank K. Wilhelm. "Characterization of decohering quantum systems: Machine learning approach." *Physical Review A* 93.1 (2016): 012122.

Measurement Protocol for the Entanglement Spectrum of Cold Atoms

Alireza Seif
Joint Quantum Institute, University of Maryland/NIST

Abstract: Entanglement, and, in particular, the entanglement spectrum, plays a major role in characterizing many-body quantum systems. While there has been a surge of theoretical works on the subject, no experimental measurement has been performed to date because of the lack of an implementable measurement scheme. Here, we propose a measurement protocol to access the entanglement spectrum of many-body states in experiments with cold atoms in optical lattices. Our scheme effectively performs a Ramsey spectroscopy of the entanglement Hamiltonian and is based on the ability to produce several copies of the state under investigation, together with the possibility to perform a global swap gate between two copies conditioned on the state of an auxiliary qubit. We show how the required conditional swap gate can be implemented with cold atoms, either by using Rydberg interactions or coupling the atoms to a cavity mode. We illustrate these ideas on a simple (extended) Bose-Hubbard model where such a measurement protocol reveals topological features of the Haldane phase.

Reference: Pichler, Hannes, et al. "Measurement Protocol for the Entanglement Spectrum of Cold Atoms." Physical Review X 6.4 (2016): 041033.

AUTHOR INDEX